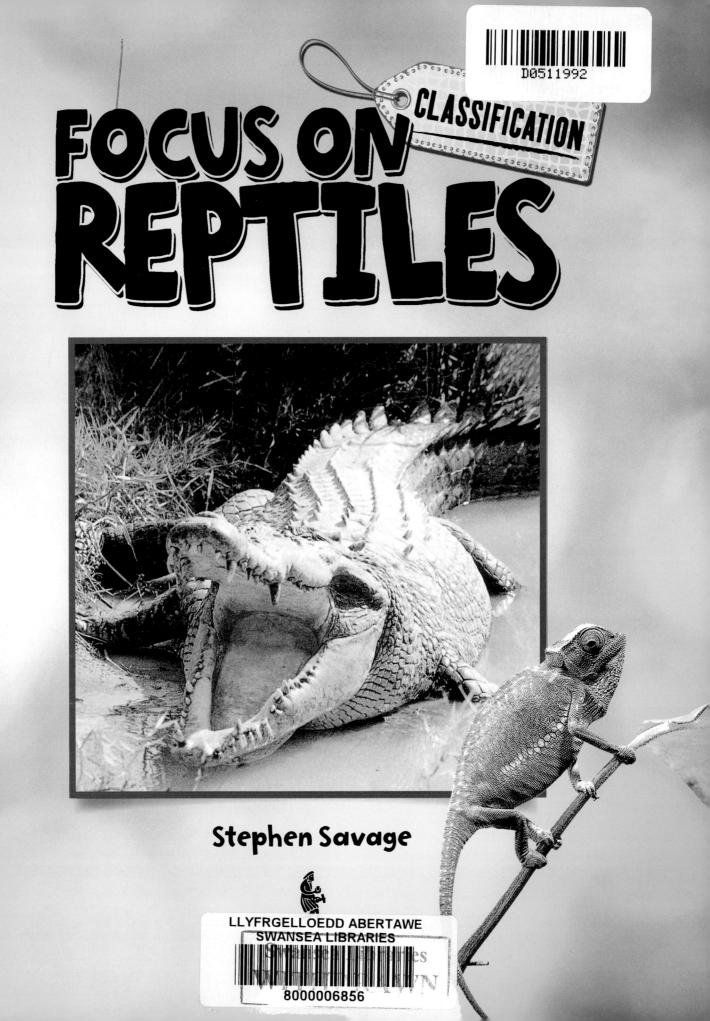

FOCUS ON

CLASSIFICATION

REPTILES

Stephen Savage

This edition published in 2014 by Wayland
Copyright © Wayland 2014

Wayland
338 Euston Road
London NW1 3BH

Wayland Australia
Level 17/207 Kent Street
Sydney, NSW 2000

Editor: Carron Brown
Designer: Alyssa Peacock

Dewey number: 597.9-dc23

ISBN 978 0 7502 8075 4

Printed in China

10 9 8 7 6 5 4 3 2 1

Picture acknowledgements:
Bruce Coleman 17(t), Shutterstock cover picture, /Rod Williams 7(t), /Joe McDonald 7(b),
/Gunter Ziesler 9, /Fred Bruemmer 14, /MPL Fogden 15(t) and 17(b), /Allan Potts 20, /Alain Compost 22(b)
and contents page, /Shaun Wilkinson 25(b), /Gerald Cubitt 26, /Joe McDonald 27(b); Frank Lane Picture
Agency /Martin Withers 6, /Brake/Sunset 23; NHPA Anthony Bannister 10(t), /Philippa Scott 10(b), /Jany
Sauvanet 11(b), /A.N.T. 13(t) and title page, /Martin Harvey 13(b), /Stephen Dalton 15(b) and title page inset,
/Daniel Heuclin 16. /Anthony Bannister 18(t), Stephen Dalton 18(b), /Martin Harvey 19, /Daniel Heuclin
21(t), /Karl Switak 21(b), /Daniel Heuclin 24, /Joe Blossom 2S(t); Oxford Scientic Films /Richard Packwood
cover(inset), Mark Jones 4, IStephen Downer 8-9, /David Dennis 12, /J. Gutierrez Acha 22(r), /Howard Hall
27(t); Wayland Picture Library/Julia Waterlow 11(t).

First published in 1999 by Wayland

Wayland is a division of Hachette Children's Books, an Hachette UK company.
www.hachette.co.uk

Contents

What a difference!

There are four types of reptiles. They are: lizards and snakes, crocodilians (which include crocodiles and alligators), turtles and tortoises, and tuataras. These four types of reptiles are very different in shape, size and colour.

Although a tortoise looks very different from a snake, both have similar reptile features.

← The giant tortoise feeds on plants and fruit. It can grow as long as 1.2 metres (4 feet) and may weigh more than 250 kilograms (550 pounds).

↑ The rattlesnake shakes its tail when it is alarmed, sounding a rattle that warns attackers to keep away from its poisonous bite.

REPTILE CHARACTERISTICS

* Reptiles have dry, scaly skin. In some cases the skin forms a hard shell.
* Reptiles are cold-blooded. They depend on the sun to heat their bodies.
* They shed their skin as they grow.
* They have no eyelids.
* Reptiles lay waterproof eggs on land.

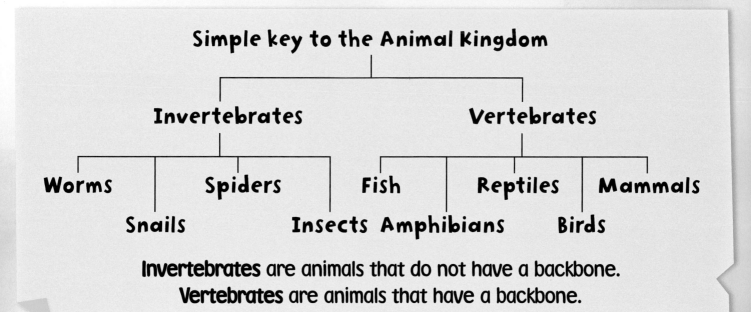

Simple key to the Animal Kingdom

Invertebrates   **Vertebrates**

Worms   **Spiders**   **Fish**   **Reptiles**   **Mammals**

Snails   **Insects**   **Amphibians**   **Birds**

Invertebrates are animals that do not have a backbone.
Vertebrates are animals that have a backbone.

Reptiles live in forests, grasslands, deserts, rivers and oceans. They live almost everywhere in the world except the Arctic and Antarctic, which are too cold.

An American alligator in Florida, USA.

LIVING IN DIFFERENT HABITATS

- Alligators have a flattened tail, which helps them to swim. They also have four legs for walking on land.

- The sidewinder snake raises most of its body off the scorching desert sand as it moves, so it doesn't get too hot.

↑ The snake-necked turtle lives in fresh water. Its webbed feet help it to swim.

Different reptiles have special features that help them live in these very different habitats. These features include webbed feet, long toes and coloured skin.

This tree iguana → has very long toes, which it uses to climb trees. Iguanas are a type of lizard.

Catching a meal

Most reptiles catch live animals. To do this, they need to have good eyesight and be able to move quickly. Tortoises and iguanas eat plants and fruit.

↑ A chameleon shoots out its long, sticky tongue to catch insects.

Very large snakes kill prey by coiling their bodies around it and squeezing. They swallow the prey whole. These snakes are called constrictors.

↓ A snake can dislocate its lower jaw to swallow large prey. This rock python is swallowing a deer, which is much larger than its mouth.

to 'taste' the ground or air. They follow the scent left by their prey.

Some reptiles use their bodies to lure prey. The snapping turtle attracts fish by waggin its tongue, which looks like a worm. The copperhead snake wiggles its tail to attract frogs.

A monitor lizard 'tastes' the air or signs of food or danger.

Komodo dragons are the world's largest lizards – they are up to metres (10 feet) long. They eat animals such as wild pigs and deer.

← Crocodiles and alligators do not have chewing teeth, so they swallow small prey whole. They catch large prey by pulling it underwater so it drowns, or they take large bites of flesh.

AVOIDING PREDATORS

* Turtles and tortoises have a hard shell to protect them from predators.

* Chameleons change the colour of their bodies to match their surroundings.

* If a lizard is attacked, it can shed its tail, which keeps wriggling while it escapes.

* Some reptiles are poisonous, a few have sharp spines and some pretend to be dead.

↓ This tree boa constrictor is well hidden among the leaves, ready to attack its prey.

Hot and cold

Reptiles are cold-blooded. This means that their body temperature changes with the temperature of the air or water around them.

↓ The gopher tortoise rests in its burrow during the hottest part of the day.

← This crocodile is opening its mouth wide so that the sun's heat warms the blood vessels in its mouth. This helps to warm its body.

Reptiles often sunbathe to warm up their bodies. They will move into the shade if they become too hot.

↓ This marine iguana is sunbathing before swimming in the sea. It feeds on seaweed.

Reptiles become slow or do not move at all in cold weather. Those living in countries that have cold winters will hibernate.

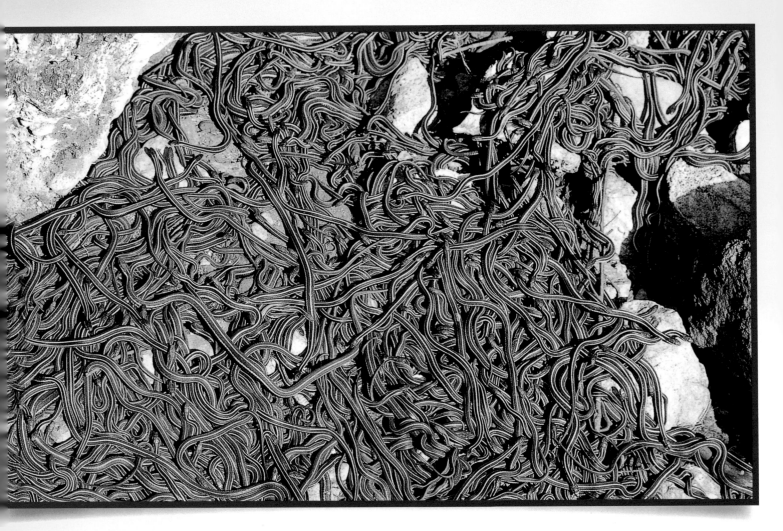

HOT AND COLD FACTS

- Reptiles' bodies are covered in scales, which keep their bodies from drying out.

- Snakes need hot air temperatures to help them digest their food.

- Thousands of male garter snakes hibernate together for warmth.

⬆ These male garter snakes have just come out of hibernation. They have slept underground through the long Canadian winter.

Some lizards can make their skin change to a darker colour. This is because animals with dark-coloured skin warm up more quickly in the sun.

↑ The Indian gharial spends the cold winter nights at the bottom of the river, where the water is a little warmer.

↑ This chameleon has become dark coloured. The sun's heat will quickly warm its body.

Getting around

Many reptiles move by walking or running on four legs. Some have no legs and move by slithering their bodies along the ground.

↓ The gila monster is a slow-moving lizard. It lives in the rocky deserts of North America.

↑ The green turtle swims with the help of flippers. It can swim underwater for four hours without coming up for air.

Some reptiles are able to climb trees in search of food. Others spend most of their life in water.

The house gecko's special toe → pads help it to climb up walls and along ceilings.

Snakes grip the ground with large scales on their undersides. These scales help them to move along the ground or to climb trees.

When escaping from danger, some reptiles run on two legs. Sometimes this can be faster than running on four legs.

⬆ The poisonous black mamba is the fastest snake. It can move at 23 kilometres (14 miles) per hour.

⬇ The basilisk lizard can run for short distances on the surface of water. Basilisk lizards are sometimes called Jesus lizards.

MOVING AROUND

- The giant tortoise moves very slowly. Its fastest speed is 6.4 kilometres (4 miles) per hour, which is the average person's walking speed.

- The flying gecko has webbed feet and skin flaps that help it glide from tree to tree.

This boa constrictor is climbing a tree, using its large scales to grip the tree trunk.

Reptile young

Some female reptiles lay eggs. They always bury their eggs on land. Even sea turtles return to land to lay their eggs.

⬇ Female Pacific green turtles come ashore to lay their eggs and bury them on a sandy beach. Most then return to the sea.

Not all reptiles lay eggs. Many give birth to live young, which develop inside the female parent.

← A newborn baby boa constrictor, still in its egg sac.

↓ A female tree python coils around her eggs to keep them warm. Can you see one of the baby pythons just hatching?

↑ These ribbon snakes are courting.
They coil around each other and their
'courtship dance' may last an hour.

Most reptiles do not look after their eggs or their young. They bury the eggs in the ground and leave the young to hatch.

REPTILE EGGS

- Reptile eggs have tough, leathery shells.

- Young snakes have a special egg tooth to cut their way out of the egg when they hatch.

- Baby crocodiles call to their mother when they are ready to hatch.

↓ A mother crocodile looks after her newly hatched young carefully. She carries them to the water in her mouth and will stay with them for a few weeks.

TAKING CARE OF REPTILES

- Pet reptiles must be kept in a tank with air holes and a tight-fitting lid.

- Use a special 'sunlight' bulb to light the tank.

- Use a heat lamp to keep the tank at the correct temperature.

- Put tree branches, bark and rocks in the tank.

- Remember that some reptiles need to feed on live animals.

Many people are afraid of snakes, even though most are harmless. Keeping a snake as a pet can be rewarding, and can show people that if treated carefully, they are not frightening.

This child is handling a pet python. →

↓ Tortoises have always been popular pets. They usually need a warm, dark place in which to hibernate during the winter.

Unusual reptiles

↑ Tuataras have lived on earth for millions of years, since before the time of the dinosaurs. They now live only in New Zealand.

Many species of reptiles have existed since the time of the dinosaurs, over 65 million years ago. Some have changed very little since. They have strangely shaped features.

STRANGE FACTS

- The Marion's tortoise is thought to be the world's longest-lived animal. It can live for 150 years.

- The slow worm is a type of lizard that has no legs. It is often mistaken for a snake.

↑ This poisonous sea snake lives in the sea, where it feeds on fish.

Reptiles adapted to live at a time when there were a lot of large predators, most of which were also reptiles. They may have needed to protect themselves by looking very fierce.

A chameleon can swivel its eyes so that one → eye looks forwards while the other looks behind.

Scale of reptiles

Human   Rock python   Giant tortoise   Diamond back rattlesnake   Florida alligator   Sidewinder snake   Green iguana

Human   Boa constrictor   Komodo dragon   Monitor lizard   Nile crocodile   Marine iguana

Human   Indian gharial   Black mamba   Pacific green turtle   Green tree python   Olive sea snake

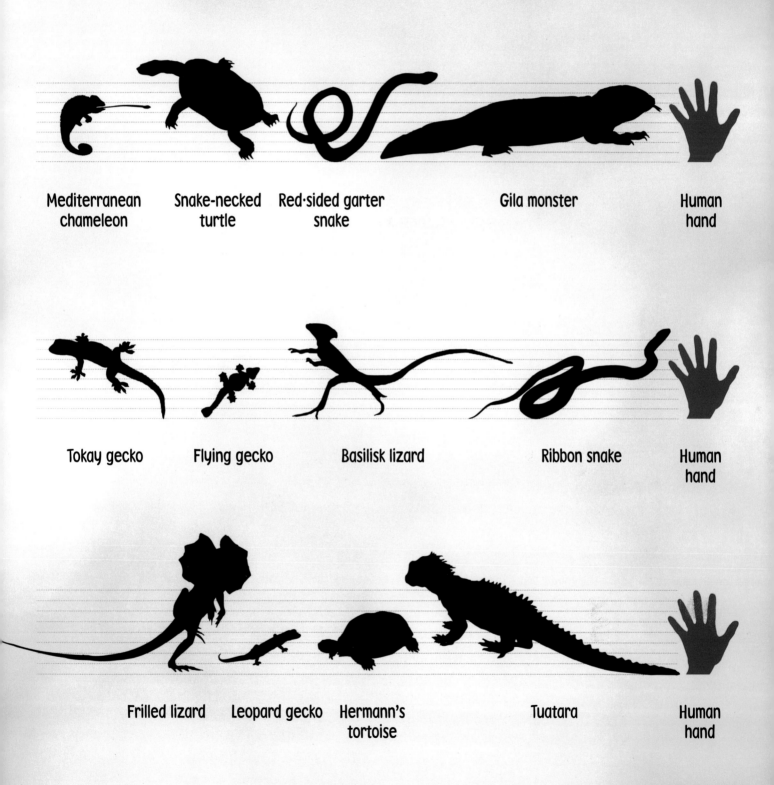

Mediterranean
chameleon

Snake-necked
turtle

Red-sided garter
snake

Gila monster

Human
hand

Tokay gecko

Flying gecko

Basilisk lizard

Ribbon snake

Human
hand

Frilled lizard

Leopard gecko

Hermann's
tortoise

Tuatara

Human
hand

Topic web

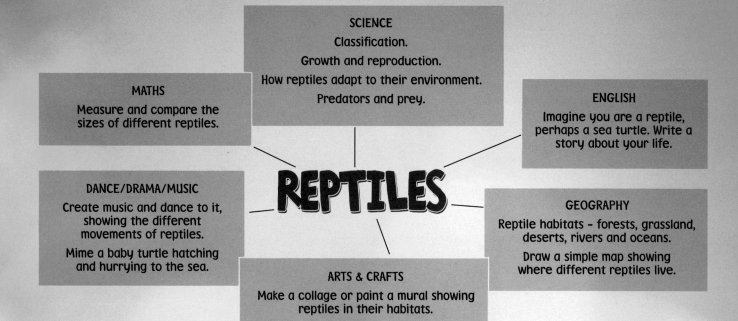

SCIENCE
Classification.
Growth and reproduction.
How reptiles adapt to their environment.
Predators and prey.

MATHS
Measure and compare the sizes of different reptiles.

ENGLISH
Imagine you are a reptile, perhaps a sea turtle. Write a story about your life.

DANCE/DRAMA/MUSIC
Create music and dance to it, showing the different movements of reptiles.
Mime a baby turtle hatching and hurrying to the sea.

REPTILES

GEOGRAPHY
Reptile habitats - forests, grassland, deserts, rivers and oceans.
Draw a simple map showing where different reptiles live.

ARTS & CRAFTS
Make a collage or paint a mural showing reptiles in their habitats.

Activities

Science Make a simple key that shows the important differences between the main reptile types: snake, lizard, turtle, tortoise and crocodile/alligator. Divide them first into land and water reptiles and then list the main characteristics of each. For instance, legs or no legs, shells, etc.

English Sea turtles have lived in the oceans since the time of the dinosaurs and have

Geography Make a simple map of the world and mark the major habitats: deserts, rainforests, grasslands, oceans, and big rivers and lakes. Find out the name of at least one reptile that lives in each of these habitats. Draw a simple picture of each, cut them out, and stick them on the map.

Arts & Crafts Make a model of a chosen reptile using modelling clay. Paint it in natural colours. It may have colours that act as camouflage, or

Glossary

Antarctic The region at or around the South Pole.

Arctic The region at or around the North Pole.

Blood vessels Veins and arteries that carry blood around the body.

Burrow A hole dug in the ground by some animals for shelter.

Constrictors Very large snakes, including boa constrictors and pythons, that kill their prey by coiling their bodies around it and squeezing.

Courtship dance A dance performed to attract a mate.

Crocodilians Crocodiles, alligators, gharials and caymans.

Dinosaurs Huge, four-footed reptiles that lived over 65 million years ago.

Dislocate To displace bones; in the case of snakes, to temporarily move the jawbones out of their joints.

Hibernate To spend the winter in an inactive state resembling sleep.

Predators Animals that hunt others for food.

Prey Animals that are hunted and killed for food.

Scales Thin bony or horny overlapping plates protecting the skin of reptiles and fish.

Finding out more

Books to read

Explorers: Reptiles by Claire Llewellyn (Kingfisher, 2013)

Eye Wonder: Reptiles by Simon Holland (Dorling Kindersley, 2013)

I Wonder Why Snakes Shed Their Skin by Amanda O'Neill (Kingfisher, 2011)

Really Weird Animals: Snakes and Lizards by Claire Hibbert (Franklin Watts, 2012)

Websites

BBC
www.bbc.co.uk/nature/life/Reptile/by/rank/all
Explore the BBC's collection of reptiles, with videos showing them in action.

Wildlife Watch
www.wildlifewatch.org.uk/explore-wildlife/animals/reptiles
Check out this website to discover which reptiles live in the UK.

Index

Page numbers in **bold** refer to photographs.